Welcome! Welcome! Look what's new,
Rocky Mountain Snowflakes 2!
All new patterns, none the same,
Every snowflake has a name.

Every snowflake has a rhyme
They brighten up the wintertime
Every snowflake has a reason
A good thing for the cold, cold season.

All are anatomically correct,
With exactly six points, as one would expect.
It's fun to fold, then snip, snip, snip
Cut a little here, here a little clip.

Decorate the windows, put them on the door,
Hang them from the ceiling, and then, cut some more.
Share your snowflakes with a friend,
They're fun to keep and fun to send.

Avalanches, blizzards, moguls and snowbanks,
Have fun with these flakes
And many thanks!

Rocky Mountain Snowflakes 2
Copyright 1998 by Debra Bangert Bonzek
All rights reserved.

Rocky Mountain Snowflakes 2 may not be reproduced in whole or in part by any method without written permission from the publisher.

Dream Bee Publications
3325 C 1/2 Road
Palisade, Colorado 81526
(970)434-7501

ISBN 0-9661572-1-4

The original book, **Rocky Mountain Snowflakes**, shown above, includes the twenty patterns at the right. The first book also includes five pages which are blank, for those who like to design their own flakes.

The special paper in both books was chosen for its pure whiteness and because it is thin and folds easily. Ask for this first book where **Rocky Mountain Snowflakes 2** is sold.

Rocky Mountain Snowflakes
These are the snowflakes found in the first book

Left to right, top to bottom: butterflies, candles, bells and the ice queen reindeer, daisy, six geese a layin' and snowmen, praying angles, ballerinas, unicorns and the Santa Fe, crystal, holy stars, Santa and flying angels, ice web, nineteen stars, bare branches and snow buntings.

The original *Rocky Mountain Snowflakes*:
ISBN 0-9661572-0-6

Instructions and Tips

Instructions are printed on every snowflake. *Rocky Mountain Snowflakes* are designed so that all the printing cuts away.

If written instructions seem confusing, try following the diagrams, just 1,2,3,4,cut!

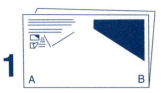

1 When folding the snowflake in half (A and B, *through center of the dot*) it may help to even up the *edges* of the paper as you fold it in half. Try to fold the center dot exactly in half. Turn the paper any way that's comfortable for you, but as you begin the second fold, make sure the directions are facing you.

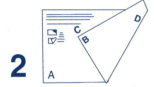

2 When making the second fold (diagonally B up to C), it helps to hold the center dot with your fingernail. If you keep the dot at the very *center point*, the finished flake will be more even. You can crease the paper *gently* for more accurate folding. Be careful, too much creasing may cause it to tear.

3 As you make the third fold (A up to D) try to keep the edges even. You are folding the paper so that you will be cutting through *12 layers!* As you fold each time, the paper becomes harder to keep even because of all the layers and folds. Just get as close as you can. Try to keep the dotted lines right on the folds, or as close to the fold as you can.

4 Finally, fold in half (along the F) and you're ready to cut. Please note that some of the patterns have so much white along the F fold that there is no place to put the dotted line. *Just fold it in half as shown here.*

cut! Sometimes it's easier to turn the paper as you cut, rather that moving the scissors.

Cutting Tips

Use sharp scissors. Remember, it's twelve layers thick! **Cut out the shaded area.** It helps to cut along the solid lines, but sometimes you may want to divide the shaded area up and cut it out in pieces.

You must not cut into the white area. The white area *is the snowflake,* and any cuts into the flake can make it fall apart. If you find a shaded area too difficult to cut out, try leaving it and see if you like the flake anyway! Remember, what you may think is a "mistake" may turn out to be the most creative and interesting flake of all!

The Shepherdess is tricky to cut. If the folding is off a little, as shown here, the snowflake won't be symmetrical. The more exact the fold, the better.

Notice the extra lines in the shaded areas. If you *cut out sections of the shaded area* as these lines suggest, it is easier than trying to cut the whole piece at once.

In areas such as the cats whiskers, straight cuts add detail.

Teddy Bears

Six little teddy bears just can't wait to get cut out and hibernate

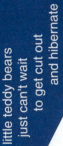

1. Tear this page out of the book along the perforation. Fold step by step *keeping the instructions facing you.* Everything cuts away leaving only the finished snowflake.
2. Fold paper up and back along the dotted lines **A** and **B** (at the left and right edges). Crease gently.
3. Fold **B** up to dotted line **C**. Crease along dotted line **D**.
4. Fold **A** up to dotted line **D**. Crease along fold **E**.
5. Fold in half and crease along dotted line **F**.
6. Cut off instructions along the trim line.
7. Cut along all blue lines to remove blue area. Don't cut into the white! Unfold gently.

The Alleycats

Alleycats caught in a sneaky pose, cut slits for whiskers under the nose

1. Tear this page out of the book along the perforation. Fold step by step *keeping the instructions facing you.* Everything cuts away leaving only the finished snowflake.

2. Fold paper up and back along the dotted lines **A** and **B** (at the left and right edges). Crease gently.

(B) Fold edge of paper up to here ▼

3. Fold **B** up to dotted line **C**. Crease along dotted line **D**.

4. Fold **A** up to dotted line **D**. Crease along fold **E**.
▼ Edge of paper here.

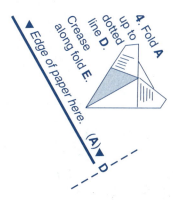

5. Fold in half and crease along dotted line **F**.
6. Cut off instructions along the trim line.
7. Cut along all blue lines to remove blue area; cut into solids. Don't cut into the white!
Unfold gently.

trim line

Unicycle Clowns

It just seems natural for a clown to peddle around and around and around

1. Tear this page out of the book along the perforation. Fold step by step *keeping the instructions facing you*. Everything cuts away leaving only the finished snowflake.

2. Fold paper up and back along the dotted lines **A** and **B** (at the left and right edges). Crease gently.

3. Fold **B** up to dotted line **C**. Crease along dotted line **D**.

▶ (**B**) Fold edge of paper up to here

4. Fold **A** up to dotted line **D**. Crease along fold **E**.

▶ Edge of paper here. (**A**) ▶

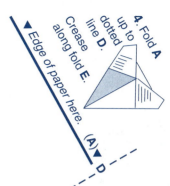

5. Fold in half and crease along dotted line **F**.

6. Cut off instructions along the trim line.

7. Cut along solid blue lines to remove all blue areas. Don't cut into the white! Unfold gently.

Cut the tire, then the spokes

trim line

Aspen Leaves

Deep in December snip into September and follow, follow the lines

1. Tear this page out of the book along the perforation. Fold step by step *keeping the instructions facing you.* Everything cuts away leaving only the finished snowflake.
2. Fold paper up and back along the dotted lines **A** and **B** (at the left and right edges). Crease gently.

▼ (**B**) Fold edge of paper up to here

3. Fold **B** up to dotted line **C**. Crease along dotted line **D**.

4. Fold **A** up to dotted line **D**. Crease along fold **E**.

▼ Edge of paper here.

5. Fold in half and crease along dotted line **F**.
6. Cut off instructions along the trim line.
7. Cut along all blue lines to remove blue area. Don't cut into the white! Unfold gently.

trim line

Snow Dance

Six Native American girls dance with the winter as the snow swirls and twirls

1. Tear this page out of the book along the perforation. Fold step by step *keeping the instructions facing you*. Everything cuts away leaving only the finished snowflake.
2. Fold paper up and back along the dotted lines **A** and **B** (at the left and right edges). Crease gently.
3. Fold **B** up to dotted line **C**. Crease along dotted line **D**.

4. Fold **A** up to dotted line **D**. Crease along fold **E**.
5. Fold in half and crease along dotted line **F**.
6. Cut off instructions along the trim line.
7. Cut along all blue lines to remove blue solids. Cut into tan area. Don't cut into the white! Unfold gently.

Hot Dog Skier

This mogul ridin' cowboy knows it's saddle up those skis when the cold wind blows

4. Fold **A** up to dotted line **D**. Crease along fold **E**.
▼ Edge of paper here.

1. Tear this page out of the book along the perforation. Fold step by step *keeping the instructions facing you*. Everything cuts away leaving only the finished snowflake.
2. Fold paper up and back along the dotted lines **A** and **B** (at the left and right edges). Crease gently.
3. Fold **B** up to dotted line **C**. Crease along dotted line **D**.
5. Fold in half and crease along dotted line **F**.
6. Cut off instructions along the trim line
7. Cut along line **E** cut into area to remove all blue solid. Don't cut into area of white! Unfold gently.

Snow Lizards

Little dreamers
noses to noses
waiting to thaw
when winter
closes

1. Tear this page out of the book along the perforation. Fold step by step *keeping the instructions facing you.* Everything cuts away leaving only the finished snowflake. ▼
2. Fold paper up and back along the dotted lines **A** and **B** (at the left and right edges). Crease gently.
3. Fold **B** up to dotted line **C**. Crease along dotted line **D**.

4. Fold **A** up to dotted line **D**. Crease along fold **E**. ▼ Edge of paper here.

5. Fold in half and crease along dotted line **F**.
6. Cut off instructions along the trim line.
7. Cut along all blue lines to remove blue solid area. Don't cut into the white! Unfold gently.

trim line

Dinosaurs

These dinosaurs just do not think that they look a bit extinct

1. Tear this page out of the book along the perforation. Fold step by step *keeping the instructions facing you.* Everything cuts away leaving only the finished snowflake.
2. Fold paper up and back along the dotted lines **A** and **B** (at the left and right edges). Crease gently.

▶ (**B**) Fold edge of paper up to here

3. Fold **B** up to dotted line **C**. Crease along dotted line **D**.

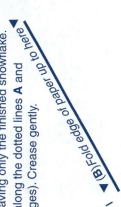

4. Fold **A** up to dotted line **D**. Crease along fold **E**.

▶ Edge of paper here. (**A**) ▶

5. Fold in half along dotted line **F**.
6. Cut off instructions along the trim line.
7. Cut along all lines. Cut into solid blue area to remove **E** all lines. Don't cut into the white!

Candles Aflame

Eighteen candles burning bright
bring some light
to a winter night

1. Tear this page out of the book along the perforation. Fold step by step *keeping the instructions facing you*. Everything cuts away leaving only the finished snowflake.
2. Fold paper up and back along the dotted lines **A** and **B** (at the left and right edges). Crease gently.
3. Fold **B** up to dotted line **C**. Crease along dotted line **D**.

4. Fold **A** up to dotted line **D**. Crease along fold **E**.

5. Fold in half and crease along dotted line **F**.
6. Cut off the instructions.
7. Cut along the trim line; cut into all blue lines to remove blue area. Don't cut into the white! Unfold gently.

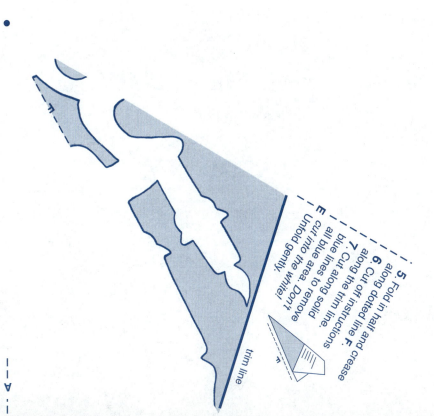

Giraffe Ring

Look closely next time it snows, look for giraffes nose to nose

1. Tear this page out of the book along the perforation. Fold step by step *keeping the instructions facing you.* Everything cuts away leaving only the finished snowflake.
2. Fold paper up and back along the dotted lines **A** and **B** (at the left and right edges). Crease gently.
3. Fold **B** up to dotted line **C**. Crease along dotted line **D**.
4. Fold **A** up to dotted line **D**. Crease along fold **E**.
5. Fold in half along dotted line **F**.
6. Cut off instructions along the trim line.
7. Cut along solid blue lines to remove all blue area. Don't cut into the white! Unfold gently.

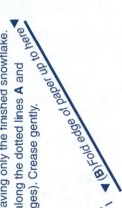
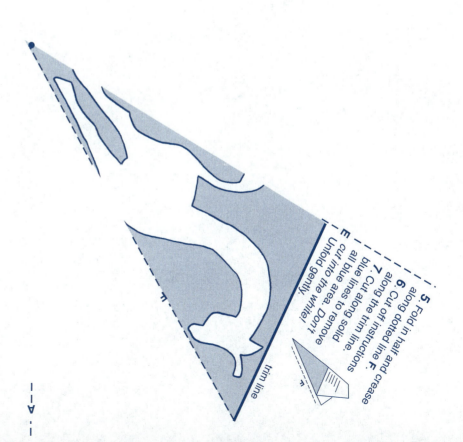

Rose Wreath

These roses are white snowflakes are too, so, why not combine the two?

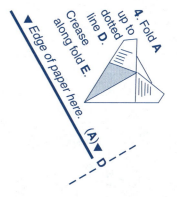

4. Fold **A** up to dotted line **D**. Crease along fold **E**.
▶ Edge of paper here. (A)▶ D

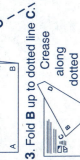

1. Tear this page out of the book along the perforation. Fold step by step *keeping the instructions facing you.* Everything cuts away leaving only the finished snowflake.
2. Fold paper up and back along the dotted lines **A** and **B** (at the left and right edges). Crease gently.
   ▶ (B) Fold edge of paper up to here
3. Fold **B** up to dotted line **C**. Crease along dotted line **D**.

5. Fold in half and crease along dotted line **F**.
6. Cut off instructions along the trim line.
7. Cut along all blue lines to remove all blue solid area. Don't cut into the white! Unfold gently.

carefully cut thru

trim line

Circus Elephants

Twelve pachyderms perched on a ball now you can say you've seen it all

1. Tear this page out of the book along the perforation. Fold step by step *keeping the instructions facing you*. Everything cuts away leaving only the finished snowflake.
2. Fold paper up and back along the dotted lines **A** and **B** (at the left and right edges). Crease gently.
3. Fold **B** up to dotted line **C**. Crease along dotted line **D**.

4. Fold **A** up to dotted line **D**. Crease along fold **E**.
5. Fold in half along dotted line **F**.
6. Cut off instructions along the trim line.
7. Cut along all blue lines to remove all blue areas. Don't cut into the white! Unfold gently.

Mrs. Claus

Mrs. Claus is in a whirl, Santa says she's his best girl

4. Fold **A** up to dotted line **D**. Crease along fold **E**.
▶ Edge of paper here. (A)▶ D

1. Tear this page out of the book along the perforation. Fold step by step *keeping the instructions facing you*. Everything cuts away leaving only the finished snowflake.
2. Fold paper up and back along the dotted lines **A** and **B** (at the left and right edges). Crease gently.
▶ (B) Fold edge of paper up to here ▶
3. Fold **B** up to dotted line **C**. Crease along dotted line **D**.

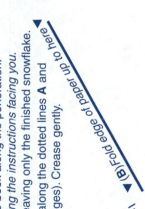

5. Fold in half and crease along dotted line **F**.
6. Cut off instructions along the trim line.
7. Cut along all blue lines to remove blue areas. Don't cut into area **E**! Unfold gently.

trim line

Ice Castles

Storm the walls
swim the moat
cut it out
so you can
gloat

1. Tear this page out of the book along the perforation. Fold step by step *keeping the instructions facing you*. Everything cuts away leaving only the finished snowflake.
2. Fold paper up and back along the dotted lines **A** and **B** (at the left and right edges). Crease gently.

▼ **(B)** Fold edge of paper up to here

3. Fold **B** up to dotted line **C**. Crease along dotted line **D**.

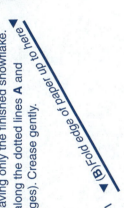

4. Fold **A** up to dotted line **D**. Crease along fold **E**.

▼ Edge of paper here.

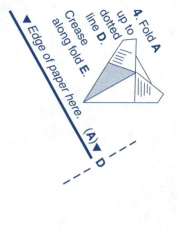

5. Fold in half and crease along dotted line **F**.
6. Cut off instructions along the trim line.
7. Cut along all blue lines to remove blue areas. Don't cut into the white! Unfold gently.

trim line

The Shepherdess

When they have their shepherdess near the little lambs have nothing to fear

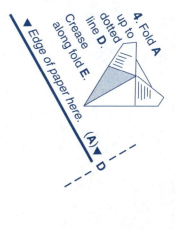

4. Fold **A** up to dotted line **D**. Crease along fold **E**.
▼ Edge of paper here. (**A**) ▼ ---- **D**

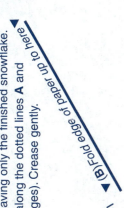

5. Fold in half and crease along dotted line **F**.
6. Cut off instructions along the trim line.
7. Cut along all blue lines to remove all blue area. Don't cut into the white!
E — Unfold gently.

trim line

▼ (**B**) Fold edge of paper up to here

3. Fold **B** up to dotted line **C**. Crease along dotted line **D**.

1. Tear this page out of the book along the perforation. Fold step by step *keeping the instructions facing you.* Everything cuts away leaving only the finished snowflake.
2. Fold paper up and back along the dotted lines **A** and **B** (at the left and right edges). Crease gently.

Racoons

Crafty racoons, here are six
Cut them out
Just for kicks

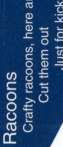

1. Tear this page out of the book along the perforation. Fold step by step *keeping the instructions facing you*. Everything cuts away leaving only the finished snowflake.
2. Fold paper up and back along the dotted lines **A** and **B** (at the left and right edges). Crease gently.
3. Fold **B** up to dotted line **C**. Crease along dotted line **D**.
4. Fold **A** up to dotted line **D**. Crease along fold **E**.
5. Fold in half and crease along dotted line **F**.
6. Cut off instructions along the trim line.
7. Cut along solid blue lines to remove all blue area. Don't cut into the white!

Swords & Shields

For truth and justice
ready to fight
perfect for a
winter knight

- B -

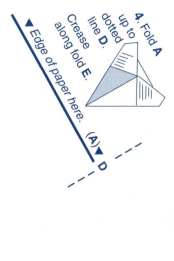

4. Fold **A** up to dotted line **D**. Crease along fold **E**.
▶ Edge of paper here. (A)▶ --- D

1. Tear this page out of the book along the perforation. Fold step by step *keeping the instructions facing you*. Everything cuts away leaving only the finished snowflake.

2. Fold paper up and back along the dotted lines **A** and **B** (at the left and right edges). Crease gently.
▶(B) Fold edge of paper up to here

3. Fold **B** up to dotted line **C**. Crease along dotted line **D**.

5. Fold in half and crease along dotted line **F**.
6. Cut off instructions along the trim line.
7. Cut along solid blue lines to remove all blue area. *Don't cut into the white!*

trim line

- A -

Porcupines

Snuggled for warmth in the cold, cold weather, these porcupines know how to stick together.

1. Tear this page out of the book along the perforation. Fold step by step *keeping the instructions facing you.* Everything cuts away leaving only the finished snowflake.

2. Fold paper up and back along the dotted lines **A** and **B** (at the left and right edges). Crease gently.

▶ (**B**) Fold edge of paper up to here

3. Fold **B** up to dotted line **C**. Crease along dotted line **D**.

4. Fold **A** up to dotted line **D**. Crease along fold **E**.

▶ Edge of paper here. (**A**) ▶

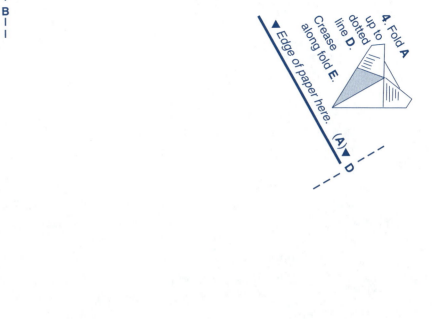

5. Fold in half and crease along dotted line **F**.
6. Cut off instructions along the trim line.
7. Cut along all lines to remove blue area. Don't cut into the white! Unfold gently.

trim line

Snow Bunnies

Here they wait
beside their door,
they don't mind
if it snows
some
more

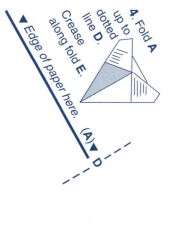

4. Fold **A** up to dotted line **D**. Crease along fold **E**.
▶ Edge of paper here. (A)▶ D

1. Tear this page out of the book along the perforation. Fold step by step *keeping the instructions facing you*. Everything cuts away leaving only the finished snowflake.
2. Fold paper up and back along the dotted lines **A** and **B** (at the left and right edges). Crease gently.
▶ (B) Fold edge of paper up to here ▶
3. Fold **B** up to dotted line **C**. Crease along dotted line **D**.

5. Fold in half and crease along dotted line **F**.
6. Cut off instructions along the trim line.
7. Cut along solid blue lines to remove all blue area. Don't cut into the white! Unfold gently.

Turtles

Get in the swim with turtles from the sea, cutting this flake is easy!

1. Tear this page out of the book along the perforation. Fold step by step *keeping the instructions facing you.* Everything cuts away leaving only the finished snowflake.
2. Fold paper up and back along the dotted lines **A** and **B** (at the left and right edges). Crease gently.
   ▼ **(B)** Fold edge of paper up to here
3. Fold **B** up to dotted line **C**. Crease along dotted line **D**.

4. Fold **A** up to dotted line **D**. Crease along fold **E**.
   ▼ Edge of paper here. (A)▼

5. Fold in half and crease along dotted line **F**.
6. Cut off instructions along the solid trim line.
7. Cut along all blue lines to remove the blue area. Don't cut into the white!

trim line

Forty-eight Legs

If you love entomology
You'll like this
flake's ecology

1. Tear this page out of the book along the perforation. Fold step by step *keeping the instructions facing you.* Everything cuts away leaving only the finished snowflake.
2. Fold paper up and back along the dotted lines **A** and **B** (at the left and right edges). Crease gently.

▼ (**B**) Fold edge of paper up to here

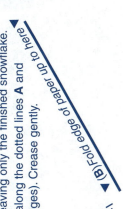

3. Fold **B** up to dotted line **C**. Crease along dotted line **D**.

4. Fold **A** up to dotted line **D**. Crease along fold **E**.

▼ Edge of paper here.

5. Fold in half and crease along dotted line **F**.
6. Cut off instructions along the trim line.
7. Cut along all blue lines to remove blue solids. *Don't cut into the white!* Unfold gently.

Siberian Tigers

Cut their stripes and cut their hats — these are ferocious cats

1. Tear this page out of the book along the perforation. Fold step by step *keeping the instructions facing you.* Everything cuts away leaving only the finished snowflake.

2. Fold paper up and back along the dotted lines **A** and **B** (at the left and right edges). Crease gently.

▼ (**B**) Fold edge of paper up to here

3. Fold **B** up to dotted line **C**. Crease along dotted line **D**.

4. Fold **A** up to dotted line **D**. Crease along fold **E**.

▼ Edge of paper here.

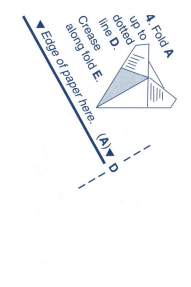

5. Fold in half and crease along dotted line **F**.

6. Cut off the instructions along the trim line.

7. Cut along all blue lines to remove blue area. Don't cut into the white!

8. Unfold gently.

trim line

Peace Doves

As you make each little dove, hope for peace and faith and love

1. Tear this page out of the book along the perforation. Fold step by step *keeping the instructions facing you.* Everything cuts away leaving only the finished snowflake.

2. Fold paper up and back along the dotted lines **A** and **B** (at the left and right edges). Crease gently.

▼ **(B)** Fold edge of paper up to here

3. Fold **B** up to dotted line **C.** Crease along dotted line **D.**

4. Fold **A** up to dotted line **D.** Crease along fold **E.**

▼ Edge of paper here.

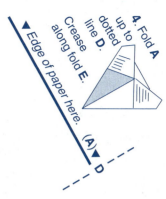

5. Fold in half and crease along dotted line **F.**

6. Cut off instructions along the trim line.

7. Cut along solid blue lines to remove all blue area. Don't cut into the white! Unfold gently.

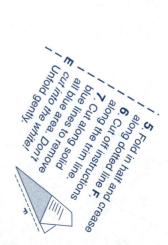

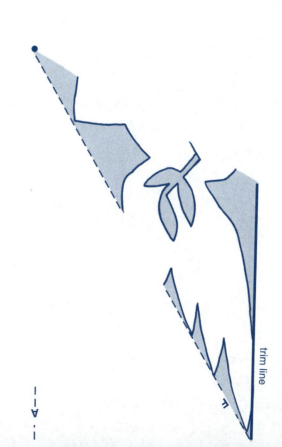

Four-string Ukulele

Cut very slowly watch each string it will help if you sing

1. Tear this page out of the book along the perforation. Fold step by step *keeping the instructions facing you.* Everything cuts away leaving only the finished snowflake.

2. Fold paper up and back along the dotted lines **A** and **B** (at the left and right edges). Crease gently.

3. Fold **B** up to dotted line **C**. Crease along dotted line **D**.

4. Fold **A** up to dotted line **D**. Crease along fold **E**.

5. Fold in half along dotted line **F**.

6. Cut off the instructions along the trim line.

7. Cut along solid blue lines to remove all blue area. Don't cut into the white! Unfold gently.

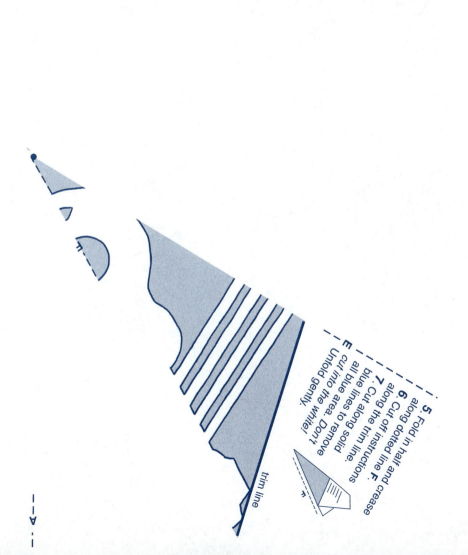

Double Luck

Horseshoes and clover win Lady Luck over and you're almost sure to win

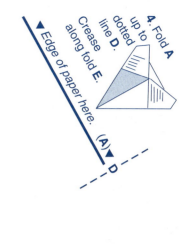

4. Fold **A** up to dotted line **D**. Crease along fold **E**.
▼ Edge of paper here.
(A)▶ D

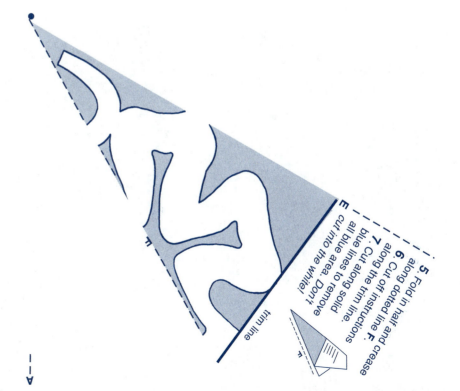

1. Tear this page out of the book along the perforation. Fold step by step *keeping the instructions facing you.* Everything cuts away leaving only the finished snowflake.
2. Fold paper up and back along the dotted lines **A** and **B** (at the left and right edges). Crease gently.
   ▼ (B) Fold edge of paper up to here.
3. Fold **B** up to dotted line **C**. Crease along dotted line **D**.
5. Fold in half along dotted line **F**.
6. Cut off the instructions along the trim line.
7. Cut along all blue lines to remove the blue area. Don't cut into the white!

trim line

Bigfoot

Be very, very, very wary
he's big, he's scary
and he's hairy

1. Tear this page out of the book along the perforation. Fold step by step *keeping the instructions facing you.* Everything cuts away leaving only the finished snowflake.
2. Fold paper up and back along the dotted lines **A** and **B** (at the left and right edges). Crease gently.
3. Fold **B** up to dotted line **C**. Crease along dotted line **D**.
4. Fold **A** up to dotted line **D**. Crease along fold **E**.
5. Fold in half and crease along dotted line **F**.
6. Cut off instructions along the trim line.
7. Cut along solid blue lines to remove all blue area. Don't cut into area marked white!